FOSSICKING AND PROSPECTING FOR GOLD.

MINING LEAFLET No. 10.

1933

IMPORTANT

Please refer to the end of this document for strict terms and conditions applying to **public fossicking** in New Zealand which must be kept at all times.

Failure to comply to these rules may result in a substantial fine or possible imprisonment.

Definition.

Shoad stone: A fragment of ore.

Auriferous quartz veins: Rock containing gold; gold-bearing

Dornie Publishing Company

Grasmere, Invercargill

www.dorniepublishing.tk

MINES DEPARTMENT NEW ZEALAND.
FOSSICKING AND PROSPECTING FOR GOLD.

Of recent months there has been greatly increased attention directed to the winning of gold in the Dominion, due partly to the extra inducement offered by the better value now being received for the metal, and partly to a large number of men turning to the work through inability to get employment at their usual callings. Already, as the result of this greater interest in the industry, there has been an appreciable increase in the output of the precious metal, and—what is more important—a considerable number of men have been enabled, by taking the work up, to make a better living than would otherwise have been possible in these times of financial stress. It is believed the men now at the work may continue to do well at it, and that even larger numbers may be employed at it, to their own advantage and that of the Dominion generally; and, with a view to helping in this direction, the Mines Department has prepared the following pages, describing the best ways of seeking for and saving gold, in the hope that they will be of real service to the workers, especially those who have had little previous experience of prospecting, and thus lead to the making of new finds, with consequent increased production and the opening-up of avenues for the useful employment of more and more men. In the strictly limited space available, it has not been possible to touch on every aspect of the work, or make the descriptions nearly as full as could be desired, but every effort has been made to include all the guidance that is most needed.

AREAS FOR PROSPECTING.

The first essential for any man taking on prospecting is to assure himself the area he proposes to examine has some promise—that is, that it is definitely gold-bearing. A large part of the Dominion is not auriferous and in this part it only means waste of time and energy to look for it. Experience has shown that the gold-bearing portions are practically confined to the Hauraki district in the North Island and the West Coast and Otago districts in the South Island. The latter districts include parts of Nelson and Marlborough Provinces. Even in these districts gold is not to be found everywhere and in the Hauraki district at least is only to be got in reefs and lodes. As most of the men now newly taking on search for it need to get a return quickly from their work, the areas mentioned in the South Island are thus the most suitable for their attention. The development of reefs is a slow process and usually a lot of expense has to be incurred before any return can. be looked for. In Western Otago, most of Westland and parts of Nelson Province nearly every creek and beach carries more or less gold and there are many old high-level watercourses and high-level gravel-beds that contain it, from which it can be won by primitive means, hence the prospector of limited means or limited experience is advised to try these districts in preference to the North Island fields. The exact locality to be investigated must be determined by circumstances. If the men going out have some definite place to go to, well, and good; if they do not know of a suitable place their best course is to write to the Inspector of Mines for the district, who will do his best to help them select one.

PROSPECTING FOR REEFS.

The searcher for new reefs usually makes a start by selecting a stream in an auriferous area, and following it up. He knows that if reefs occur up the stream, a certain amount of free gold and quartz (floater or shoad stone) will probably have been shed from them, and carried down into the stream.

He searches therefore for such indications, breaking with his pick-head any loose quartz he sees and closely examining it, or sinking small potholes in the stream, or in its banks, and panning off the material removed. If pieces of quartz are located showing gold, the prospects are bright, for this stone if followed up is almost sure to lead to the finding of the reef from which it came. All quartz does not carry gold, nor does gold-bearing quartz always show visible colour. The quartz of some reefs may be fairly rich without the gold revealing itself to the eye. It is thus advisable that the prospector should have with him a dolly-pot (the lower part of a quicksilver-bottle is mostly used for the purpose) in which to crush samples. Failing an iron vessel, empty fruit or fish tins can be made to render useful service, and many gold prospectors rely on a hard flat stone on which the samples may be crushed by using another smaller stone. If the sample is too hard to crush in this way, it may be burned in the camp fire, where it will be readily reduced to powder.

It may be said also that no particular class of quartz carries gold, the reefs of almost every locality differing in their characteristics. The prospector should therefore examine all the quartz he comes across, giving special attention to any rusty-looking samples or to pieces showing fine laminations, for these latter almost always carry values. Once it is found that any particular variety of quartz carries gold, the tracing of the shoad afterwards is simplified, for a very little practice will soon enable the prospector to recognise the right kind at a glance.

If no gold-beating shoad is found at first, there may be colours of free gold. The sinking of small pits in the stream or its banks, and the thorough examination of any rock-bars crossing the creek, will soon reveal whether or not these are present. If any colours at all are got there is hope of something better, and if it so happens that small particles of quartz are found adhering to the grains, or if the grains of gold are sharp in outline, the material has surely come from reef, perhaps at no great distance.

Once shoad or free gold has been found, the work is continued upstream, taking in any branch streams that may enter, until no further shoad-stone or colours are got. The presumption then is that somewhere up the bank on one side or the other of the stream is the reef that shed the material.

If there are rock exposures on the banks these should be at once examined, but if the country is covered with detritus from the weathering of the hills, as frequently happens, trenching and loaming must be resorted to. If by this means indications continue to be got they must be followed up until they will eventually lead to a spot where either the reef will be found to outcrop, or is buried beneath the detritus.

In the latter case trenching just above where the last shoad or colours were got should then soon reveal the reef.

If one side of the creek fails to give indications, the other must he tried. Perseverance is called for, and the prospector should never despair so long as gold is present. Fig. 1 will serve to show the method of operating in search of this kind.

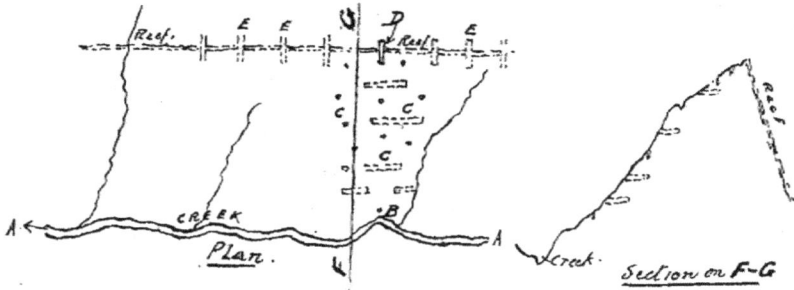

FIG 1.

A—A is the stream originally followed up, with B the point beyond which no shoad or colours were got, C. C represent the pot-holes and trenches put down in trying to trace the indications up the hillside, while D is the place where the reef was eventually located. Following the finding of the reef, the next thing is to trace it along its line by trenching as at E, E.

It by no means follows that even if a reef has been cut in several places without showing pay-values that other parts of it may not show better values. The gold often occurs in shoots in a reef, the shoots being separated one from another by runs of poor or barren quartz. Hence the necessity of cutting trenches at frequent intervals across the reef.

When extensive trenching, with careful sampling and testing of the ore, fails to reveal values, it may be taken as a general rule that there is no good purpose to be served by putting in further work on it. The only case in which this is justified is when the outcrop alone shows sign of strong leaching or oxidization, which is evidenced by the stone taking on a darkly-stained, spongy, or honeycombed appearance. Where this condition is noted, there has probably been a migration of the original surface values to a lower level, and sinking on the reef should be carried out. It may then be found that at some little distance below the surface a zone of secondary enrichment occurs, below which the reef may still carry better gold than at the outcrop. Any deeper testing of a reef in this way should be done by sinking actually on the stone rather than by driving adits, for each foot of work thus done helps directly, and in the speediest way, to prove the value of a find, and often saves much needless expense and loss of time.

PROSPECTING FOR ALLUVIAL GOLD.

The main thing in searching for alluvial gold, especially in an area not previously examined, is to have a realization of the conditions in which it is usually deposited, and, with these in mind, to carefully study the area. All alluvial gold has at one time or other been water-borne. Weathering brings about the erosion of country originally containing gold-bearing formations.

The resultant detritus is carried by rainfall into numerous small watercourses, which in turn enter larger and larger streams, down which it is swept until at least under friendlier conditions it is piled up in beaches or laid out in flats.

The gold is carried down with the rest of the material, and remains with it when it has come to rest. In the upper reaches of the steams comparatively little of the gold is left behind.

The rapid fall of the water washes the bed-rock bare, and all the gold likely to be found is in small crevices in the rock or in the shelter of large boulders. Nor does much gold remain as a rule in any portions of the streams, even at lower elevations, where there has been a narrowing of the channel.

In such places the rush of water is greater than elsewhere, and all loose material is boiled out and swept onward to lodge in more protected situations. This remark applies also to parts of a stream where the current tends to drive in under the stream banks. Where hard bars of rock cross a stream there may be catchment provided even in fairly rapid water, but in a general way it may be said that practically all deposition of the metal takes place at spots where the fall of the stream lessens, or where the force of the current is checked. This may be where side streams cutting through softer belts of the country tend to build up each flats at their junctions with other streams, where the intrusions of hard spurs of rock into a stream afford a protection to the building-up of beaches in the shelter of their lower sides, or where a stream debouches from a narrow channel into more open country.

Hence, in looking for a likely place in which to start his search, a prospector should seek such locations. The thing to determine first is if any gold at all occurs. This may perhaps be done most readily by examining any rock-bars that may be present, care being taken to thoroughly clean out any crevices.

If colours are got, the work can be extended to the beaches in the shelter of the spurs. These should be tested by trenching into the banks or sinking pot-holes. To test the larger flats below gorges it is a good plan to go back some distance from the stream, picking up with the eye what may possibly have been an earlier course of it, and sink. Well up a stream this sinking may be shallow, but lower down it will get deeper and deeper, owing to the filling-up of the old stream-bed by the material carried down through the ages. There may be a little gold right down through these beaches or flats, but the best of it will invariably be at the bottom of the gravel on the bed-rock, so every effort should be made to reach the rock, when the probability is that a satisfactory return will be got for the labour involved.

The locations mentioned are not the only ones in which gold may be sought. In the course of centuries all streams such as those that intersect the gold-bearing regions of Westland and Otago have gradually cut their channels deeper and deeper, especially in their upper courses. This often means that, apart from the gravel-beds that now line their present banks, remnants of the old beds laid down at earlier stages may still be left at various elevations up the hills on either side. These may well be gold-bearing, and careful search should always be made for them. When located, they may be quickly tested by driving in on the bed-rock.

Gold-bearing alluvial deposits may also occur in the provinces mentioned far removed from the present streams.

These may represent the courses of ancient high-level streams long since diverted in other directions, or deposits that were originally laid down at lower elevations and afterwards raised by the uplift of the land surface. Thus any accumulation of gravels, or places where the topography of the country suggests the presence in former times of an old steam channel, should always be tested, even though they are much higher than, and far from, any of the present streams, and these will be found in many places almost up to the crown of the ranges. In testing these, the essential thing is to carry the work, whether it be trenching or shaft-sinking, to bed-rock, as the best of the gold is certain to be on the latter or in the last few inches of clay, sand, or gravel resting on it. Usually such gravels will be dry, rendering this possible. If no gold is got at the first places tried, this need not discourage. Gold is rarely evenly distributed over any bottom. It usually occurs in runs or leads. One of these runs may not be far away. The dip of the bed-rock should be noted. If it is sloping, further holes should be put down to reach it on the dip aide. The deepest of the old channels are those most likely to carry the gold. Present surfaces cannot be relied on to give indication of the contour of the old bottom. The lay of the bed-rock must be the guide, and the information afforded by the trenches and shafts must be followed up till the main channels of the old surfaces are located.

FOSSICKING.

For the man not disposed to break new ground or go far from the beaten track, there is still abundance of room to do some good for himself. The sea-beaches offer opportunities for black-sanding.

Nearly every spell of heavy weather by combing down areas of beach, reveals patches of gold-bearing sand, and on almost all the creeks, even if they have been worked before, there are always chances of getting small hauls of gold. Patches of ground have been left unworked perhaps between old claim, or owing to the presence of large tree stumps, which unworked edges on creek banks may be numerous. These should be diligently sought. Even the paddocks of the old diggers are not to be despised, unless the latter happened to be Chinese. By cleaning them up again, scraping out carefully every crevice (the tools required for this are a piece of strong hoop-iron, bent near the end, for the wider crevices, and a piece of stout fencing wire similarly bent, with the end flattened to a chisel point for the narrow ones), there it always the chance of getting an appreciable amount of gold. Every crevice should be followed down as far as it extends, and every particle of clay or other material removed from it for subsequent washing. Then the small beaches should not be neglected. Even if turned over previously, these will often build up again, especially in streams liable to frequent flood, and may yield a good return. Then the streams, and all their smaller branches, may be followed up, and every bar investigated.

The crevices in these especially on their down-stream side are good traps for gold, and deserve a thorough cleaning out. Sometimes below a bar there will be deep holes which have caught gold. Those should be closely searched for, and cleaned out down to the solid rock. It may be necessary to divert the stream to do this, it build a small dam, but the work may be well repaid.

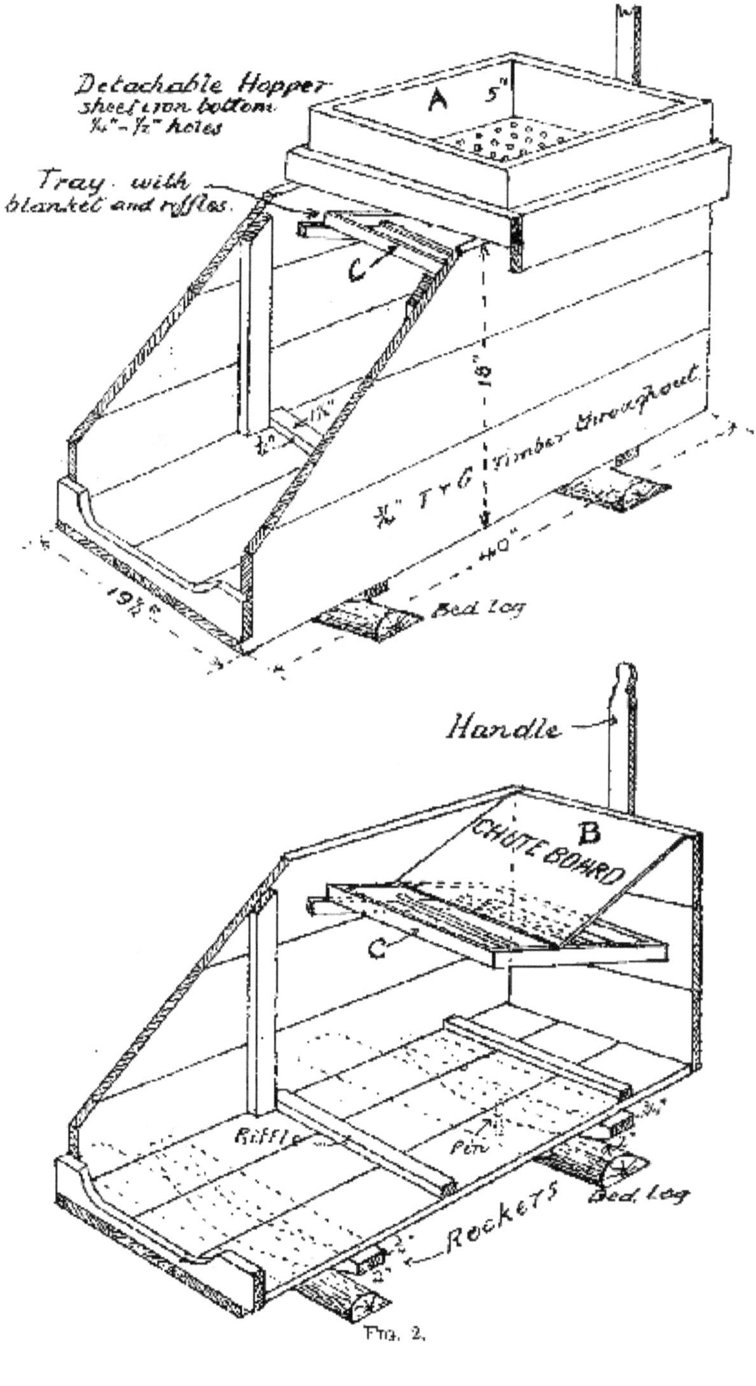

Detachable Hopper
sheet iron bottom.
¼"–½" holes

Tray with
blanket and riffles

A   5"

C

18"

¾" T & G Timber throughout

40"

19¾"

Bed Log

Handle

CHUTE BOARD

B

C

Riffle

Pin

Rockers

Bed Log

¾"

Fig. 2.

13

Then it is wonderful how large boulders in a stream will help in trapping gold. Right up the course of a gold-bearing stream wherever these are seen they should be removed. Sometimes a stout sapling lever will turn them over; at others a little explosive may be needed to effect the purpose, but most times the trouble is worth-while going to. When a boulder has been shifted, all clay adhering to its lower sides should be carefully scraped, or washed, off. Then the hole from which the boulder has come, which may be several feet in depth, should be cleaned out, every particle of clay being saved and every crevice followed down, when it will be hard luck indeed if pleasing return is not got.

METHODS OF SAVING GOLD.

In a small paper such as this it is out of the question to attempt to describe the many and complex methods of recovering gold from quartz. The subject is altogether too big to be dealt with the space available. Further, an attempt to do so would serve no useful purpose for the class of prospector for whose guidance this paper is primarily intended—viz., the man who wishes to work his own ground. There may be instances where a reef is rich enough to enable a prospector to win a good deal of gold by pounding the stone in a dolly-pot and panning or the crushed material, but this rarely happens. Occasionally, also, it may be possible by use of a small plant such as the one-stamp battery of which several types are manufactured, for the independent miner to treat his quartz in a somewhat larger way, but any technical knowledge he may require in such a case he can readily acquire.

Generally speaking, when a quartz-mine reaches the plant-erection stage it has got beyond the control of the prospector; the matter has become one for the consideration of a company; and there is then always available the services of qualified men who will be able to determine what is to be done without help from any such limited information as could be given here.

Regarding the saving of alluvial gold more may be said, for guidance in this respect is not readily available, and as many of the men who are now taking up prospecting work have not had much experience some description of the best methods in use is needed by them, and should be most helpful to them.

Cradling —The use of the cradle or rocker is widespread. It means hard work and slow work, but the cradle is most effective in operation, a careful man being able to save every particle of gold in his wash dirt, and—what is more important—the appliance costs little to make and can be constructed readily by any handy man. **Fig. 2** will serve to show the general construction, the dimensions given being those most favoured. It will be noted that on the top a detachable hopper, **A**, it provided. The material to be treated is shovelled into this, and by use of the hand the end, a rocking motion is set up.

At the same time water is poured on the dirt (usually from a "dipper" made by securing an empty fruit tin to the end of a short stick). The smallest of the wash-dirt then falls through the holes in the hopper-bottom on to the inclined chute-board **B**, down which it is directed to the tray **C**. This tray rests on cleats nailed to the inner sides of the box.

It has a wooden bottom, which is covered with canvas or bagging, across which several riffles, about ½ inch wide by ¼ in high are set. In **Fig. 2** only two of these riffles shown, but more can be used if desired. As the dirt works its way down, each of these riffles catches some of the gold, and the tray will, if properly set and worked, catch practically all that is present. Any gold or other material that is not caught by the tray riffles passes over the end of the tray and falls to the bottom of the cradle, which in turn is also usually covered with blanketing of some suitable kind and provided with several further riffles somewhat higher and wider than those of the tray. These latter serve to catch any fine gold that has escaped the tray. Beneath the cradle rockers are fixed, in the centre of which are pins that enter holes bored in the bed logs. In setting the cradle for work, it is given a fall to the front end. This is usually about 1 in. to the length of the cradle, but is really determined by the character of the gold and of the wash-dirt. If a lot of heavy sand is present, more fall is needed than with lighter material. Again, too much fall may cause loss of gold if the latter is light, flaky, or containing much impurity, while if too little fall is given the sand will pack hard behind the riffles and prevent the gold settling, What the operator has to do is to watch how his riffles are acting, and every now and then wash a dish of the tailings to see if any gold is getting away. If he finds gold in the dish, he knows the fall is either too steep or too flat, and adjusts his box till the right fall is provided.

In the illustration of the cradle only one removable tray is shown, but if desired several can be used. One is generally found sufficient, but where the gold is very fine it may be advisable to use two or even more.

When this is done, the trays are usually given reverse dips, so that the material going though zigzags down over them. The cradle is placed near a pool of water, and as near as possible to the place being worked. Any large stones are removed from the wash-dirt before it is placed in the hopper.

The rocking must be done evenly. As the dirt is washed through the screen, more is fed in. Small stones quickly collect in the hopper, which is removed from time to time so that they may be thrown out. If the work is being done in country where coarse gold occurs, the pebbles are closely examined before being thrown away, as otherwise small nuggets may be lost. After rocking for several hours, the tray also is removed and its contents washed off into a dish. If really good ground is being worked the tray should be removed every hour, and, in any case, if gold starts to show in the riffles it should at once be taken off. Every now and again, also, the sand that collects in the riffles in the bottom of the cradle should be taken out and returned to the hopper.

Should there be any clay in the wash-dirt, this must be well puddled before being put in the cradle, as the clay will "ball up" and carry gold away. The puddling may be done in a trough or tub by stirring well with a shovel or hoe, or even with the hand.

To clean up, all that is necessary is to pan off the concentrates saved by the riffles on caught on the blankets in the tray or cradle-bottom.

Cradling is recommended where only small quantities of material have to be treated, especially if the wash is fairly rich. The method is too slow for the treatment of low-grade wash. As a rule the dirt put through is only that from the bottom of the gravel on the bed-rock. If this dirt cannot be blocked out as when mining from a shaft or tunnel, it is best to resort to some method of boxing such as will now be described to save the gold, as otherwise too much poor or barren material may have to be treated.

Boxing.—This is generally resorted to where larger bodies of wash have to be dealt with than is possible with the cradle, and where a steady flow of sluicing-water is available. In New Zealand abundance of water is usually at hand. One type of box used is shown in **Fig. 3**. This is about 10 ft. long, 2 ft, wide at the head, and 1 ft. 6 in. at the tail. The side boards are 12 in. high adobe tail end and 2 ft, at the head. At the tail end vertical cleats are nailed on the inside of the side boards, against which 1 in. riffles can be set. The floor of the box, with the exception of about 2 ft. at the head end,

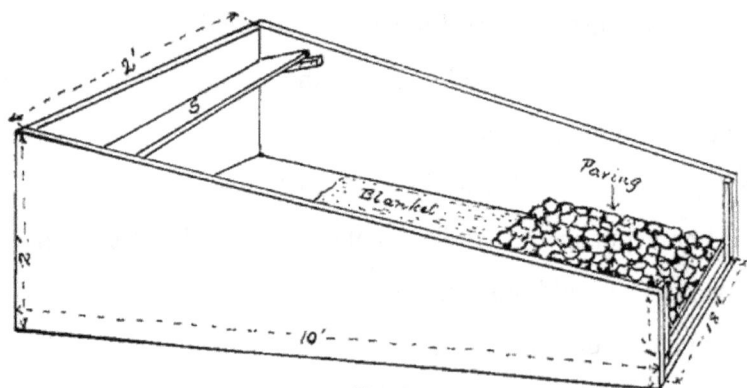

FIG. 3.

is covered with a blanket (sacking, canvas, corduroy cloth). To set the box at the right inclination, it is first put in a trench prepared for it, with the head about at the level at which the water can be led into it and with a dip of about 1 in. to its full length. A 1 in. riffle is then placed behind the cleats at the tail end, and a little water allowed to flow in. The head of the box is then raised or lowered so that the water coming over shoots straight down to where it starts to deepen owing to the banking up of the water held back by the riffle. When about 1ft. 6 in. of the upper part of the box then appears bare, or has only the merest film of water on it, when the water is thus flowing, the inclination is right, and the box is firmly packed in that position.

The box is then paved for 3 ft. or so back from the tail by packing in layers of small flat stone, the idea being that the gold will settle down between the stones and be held by the riffle. The wash to be treated is then fed in, a small barrowful at a time to the head of the box on the unblanketed part. As stones collect they are lifted out with a fork.

The work thus goes on, and the dirt can be put through as fast as it can be fed in and the stones forked out. The flow of water must be carefully regulated, no more being allowed to pass through than will just carry the tailings away.

Most of the gold will hang at the head of the box, but any that travels farther will be caught in the paving or by the riffles.

To cause the water to fall evenly into the head of the box, a splashboard, resting on cleats, as at **S**, is sometimes used, but this is not essential. A rough clean-up is usually made about every four hours, by turning off most of the water, forking out the stones from the head of the box, and gathering any gold or concentrate that is left there. At knock-off time the whole box is cleaned up. This is done by again turning most of the water, forking out the stones, and removing the gold in the head of the box, then forking out the paving, lifting the blanket and washing it in the box, and sweeping all the " fines" left down to the tail and into a dish placed there to catch them. The concentrates are then panned off, and the clean-up is completed.

As fully twenty times more dirt can be put through in this way than when using a cradle, this method of gold-saving enables much poorer gravels to be treated profitably. Another type of box that is often used, and that will give good results, is shown in **Fig. 4**. Requiring much less timber in its construction it is suitable for men working in outlying places, and where the necessity exists of shifting

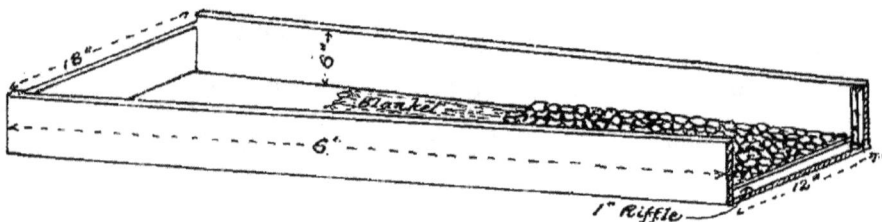

FIG. 4.

often from place to place its lightness is a recommendation for its use. This box is only 6 ft. long, 1 ft. 6 in. wide at the head, and 1 ft. at the tail, while the sides are 6 in. high. Cleats and riffles are used at the tail end as in the box previously described. The lower part of the box is also covered similarly with blanketing.

The paving-stones can also be used in a similar way, but in place of them riffles formed of slats, as at A, **Fig. 5**, or of twigs as at B can be used, while in some cases strips of timber about 2 in. thick in which numerous auger holes have been bored are sometimes preferred. In making any of these riffles, they must be framed to fit into the bottom of the box. These boxes are good gold-savers. Most of the gold will remain at the head of the box. If the gold is extremely fine, a second straight length of box can be added at the tail. This is covered with coconut fibre, plush, or some such material.

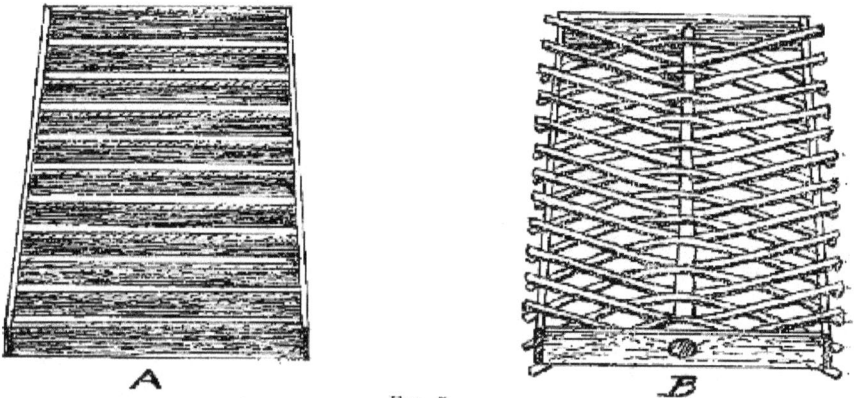

A

Fig. 5.

B

The tapered form of box is not always used—for instance, in hydraulic sluicing in a big way, straight-sided boxes up to 4 ft. or more in width, 2 ft. in depth, and many feet in length are usually employed, being paved with wooden blocks for a considerable length at the head; but the men for whose guidance these pages are meant will find all their requirements met by the use of one or other of the types of box just described.

Banjoing.—This is a method of concentrating values in gravel that is largely used for the recovering of alluvial tin in Queensland, and sometimes used for gold, but for the latter is not so effective. The "banjo" is a tapered box constructed much the same as that shown in **Fig. 3**, but much shorter, usually about 5 ft. In operating, the box is placed at an inclination with its tail close to a pool of water.

The dirt to be treated is then placed in the head, and the operator, standing in the pool, dashes dishes of water up against it with a swirling motion. This gradually washes away the sand and stone, leaving the gold or other mineral on the box-bottom, near the head. Alluvial tin can be well saved in this way, even without the use of a blanket; with the use of blanket gold may also be saved fairly well, but the prospector is advised to use one of the other methods in preference, as there is a tendency for very light or fine gold to escape. It is, moreover, back-breaking work.

Ground sluicing.—Another method in common use for recovering alluvial gold, especially from comparatively low-grade and shallow gravel deposits, is that known as "ground" or "hand" sluicing. In this, water is made to do practically all the work, and no boxes at all are used. The method is particularly applicable to gravels that are situated in such a position that a fair supply of water can be readily brought on to them, and where the bedrock is not too deep. Good fall is essential, therefore the bedrock must be at a height that will afford the necessary fall to get rid of tailings, although a good deal can be done to remedy any defect in this particular by constructing a long tail-race. A position in which this method can be resorted to is one such as illustrated in **Fig. 6**. Here a small stream is found flowing through a flat the gravels of which carry gold, though perhaps not in great quantity.

To work this, a dam is first put in at the necessary distance up the stream, say as at D, to allow the water to be diverted in a side race EE.

The water thus turned aside may be allowed to return to its old bed below the area to be worked. A tail-race is then brought up along the course of the old creek. In making this the soil need not necessarily be shovelled out.

A little of the water is allowed to run down the old bed, and, working in it, the bottom is picked up for some inches in depth. The water will carry a lot of the disturbed material away. This stirring-up of the race-bottom is repeated, several times if needed, till the bottom has been carried say a foot into the bedrock. A cross race as at F is then started between the side round the tail race. This may have to be dug to a depth of at least a few inches, after which the water from the side race in turned into it and will do the rest of the cutting, with perhaps a little lifting-up of the ground now and again with the pick. The rush of the water will cut this trench down to the bedrock, taking more or less of the sides, and carry all the gravel removed into the tail-race. To ensure that any gold in it goes to the bottom of the tail-race, the operator should go along from time to time with a shovel, and, starting at the lower end of the race,

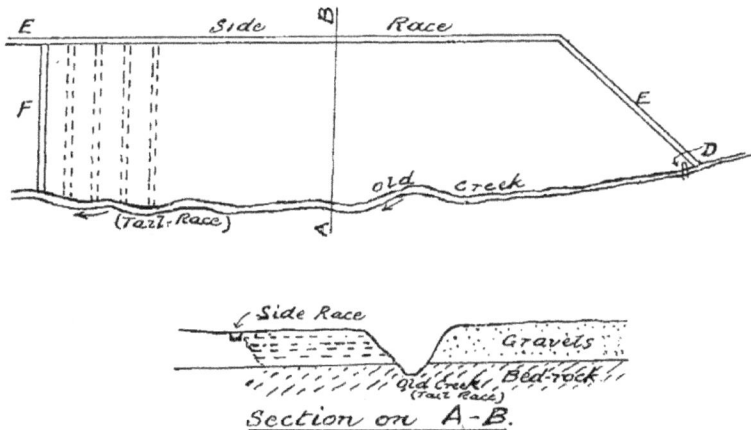

Section on A-B.

FIG. 6.

lift the material up and let it drop back again into the water. This stirring-up causes the gold to settle on the bottom. When the water has done all its work in the first cross trench, further trenches are opened in succession as shown by the dotted lines, and so the work goes or till the whole paddock has been sluiced down. Cleaning-up cannot be done often, and may not be done till the whole paddock has been dealt with. When it is entered on, the mode of procedure is to turn some of the water back into the old channel, then with the shovel, starting at the lower end of the race, to lift up the material remaining in it and drop it back again. Each stirring reduces the volume of the material, and the operation is repeated until the whole has been reduced to a few inches in the bottom of the race. The water is then turned off, and the material left carefully cleaned up for subsequent panning or streaming. When cleaning-up, the bedrock floor of the paddock must never be neglected. A good deal of gold will remain on this, and every particle of loose material must be swept down into the race. Every crevice must also be thoroughly cleaned out, otherwise a lot of gold may be lost. A large amount of material can be shifted in this method of working at very little cost, and even without much manual labour, so comparatively poor ground could be made to pay in following it.

There are other methods of working alluvial ground and saving the gold, such as dry-stacking, dredging, elevating, &c., but these are either not suitable for the Dominion or else are beyond the resources of the ordinary prospector or fossicker, consequently they call for no special description.

Some few words may be permitted as to a modified form of "hydraulicing." There may be instances where without going to great expense a prospector may be able to shift ground in quantity by the use of water under pressure, that would otherwise not pay to work. This applies to cases where water can be picked up conveniently at a moderate elevation above the ground to be worked. The use of galvanized iron pipes to bring the water to the face, and of canvas hose to which a hydrant can be fixed for breaking-down purposes, may be possible to the prospector of even moderate means. Galvanized piping does not cost very much, and the canvas hose can often be made on the ground. If a hydrant cannot be got, or is too costly, a bottle with the end removed inserted in the hose may render useful service. After being broken down, the gravel must, of course, be put through a sluice-box to save the gold.

Department of Conservation Gold Fossicking Care Code

Every person shall have the right to mine for gold in a gold fossicking area by means only of non-motorised hand held tools'—Crown Minerals Act 1991, 98 (3).

Fossick with care for the environment:

1. Be informed of all statutory regulations that govern prospecting activities in New Zealand.
2. Prospect only in the permitted area.
3. Only drive your vehicles on tracks and roads open to the public.
4. Do not remove or damage any shrubs or trees, and minimise damage to ground layer vegetation.
5. Restore the ground as you found it. Backfill any holes you dig and replace any leaf litter as it was as best you can.
6. Equipment for excavation on the land other than hand tools may not be used. Never use explosives.
7. Don't disturb, destroy, interfere with or endanger an archaeological site or place of cultural significance.

Public Fossicking Areas.

- Aorere River A
- Aorere River B
- Arrow River
- Britannia Stream
- Five Mile Creek
- Gabriels Gully
- Jones Creek (1)
- Jones Creek (2)
- Louis Creek
- Lyell Creek
- Moonlight Creek
- Nelson Creek
- New Creek
- Shamrock Creek
- Shotover River
- Slab Hut Creek
- Twelve Mile Creek
- Waiho River

For further and more accurate detail, please visit:
www.crownminerals.govt.nz/cms/minerals/gold-fossicking or
www.paydirt.co.nz and *www.findgoldnz.com* (both excellent sites for all your gold information and resources)

* Please note there is currently no "free" public fossicking areas in the North Island.

Gold Resources New Zealand

EPITHERMAL GOLD-SILVER DEPOSITS

GOLD-QUARTZ LODES

HAURAKI EPITHERMAL GOLDFIELD

GOLD PLACER FIELDS

MESOZOIC-HAAST SCHIST

PALEOZOIC OROGENIC ZONE

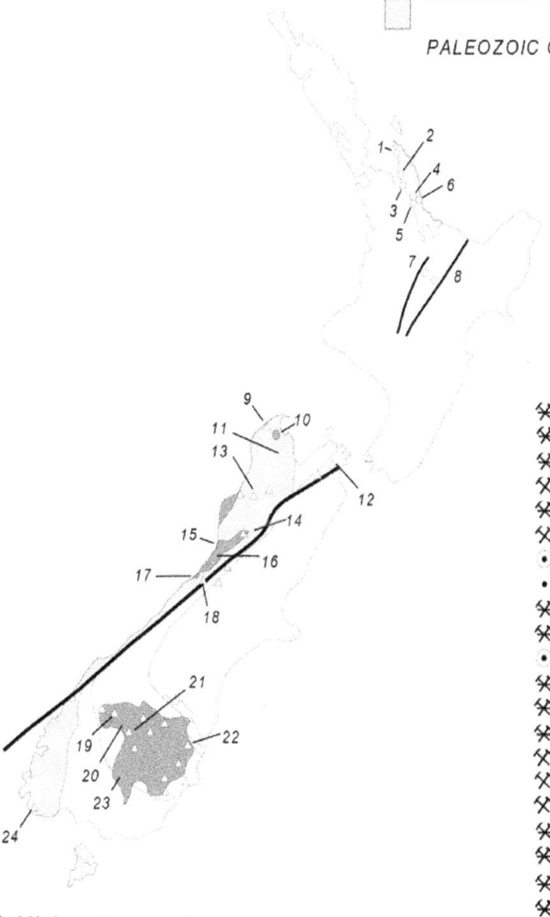

1. Coromandel
2. Monowai
3. Thames
4. Golden Cross
5. Karangahake
6. Waihi
7. Ohakuri
8. Broadlands
9. Golden Blocks
10. Collingwood
11. Sams Creek
12. Wakamarina
13. Lyell
14. Reefton
15. Grey Valley
16. Taramakau
17. Ross
18. Mt Greenland
19. Shotover
20. Kawarau
21. Cromwell
22. Macraes
23. Nokomai
24. Preservation Inlet

Mining at present

Previously mined

Significant occurrence

Occurrence

Notes